Quantum Physics for Beginners

Discover Quantum Mechanics and Physics Theories, Learn in an Easy Way Basics and Advanced Concepts and Explore a New Universe of Knowledge

By

Marc Beam

The trademarks that are used are without any consent, and the publication of the trademark is without permission or backing by the trademark owner. All trademarks and brands within this book are for clarifying purposes only and are the owned by the owners themselves, not affiliated with this document.

Table Of Contents

Introduction

- **How can you exist in two different places at the same time?**

- **How is it possible for your pet dog to be alive and dead at the same time?**

Answers to both of the questions mentioned above are possible in the mysterious world of quantum physics.

If Quantum Physics is correct, it signifies the end of physics as a science.

(Albert Einstein)

When you start looking at Quantum Physics from a beginner's perspective, this isn't really easy.

Quantum Physics or Quantum Mechanics is complex, and most of the time, professional physicists can face a difficult time trying to understand any concept of Quantum Physics or Quantum Mechanics.

As far as my opinion is concerned, there are some basic concepts and features of Quantum Physics and Quantum Mechanics around which the entire theory of Quantum revolves. If you start understanding these concepts, you will find quantum physics really simple to understand at the beginner's level.

As we already know that our entire universe is made up of billions of particles, particles can form the waves. Both of the particles and waves correlate and coexist.

Quantum physics also pushes us out of the boundaries of what we already know about reality, science, and the universe.

Quantum physics is no less than a mystery, where quantum particles can act weird and do wonderful things and can act differently to the objects we observe and experience in our daily life.

In this book, you'll learn the basics of Quantum Physics, such as:

- Quantized Properties of the Matter
- Building Blocks of our Universe
- How Quantum Physics acts in our Daily Lives
- Black Body Radiation
- Black Body Emission
- Rutherford Experiment
- Double-Slit Experiment
- Bohr's Atomic Model
- The Photo-Electric Effect
- Frank-Hertz Experiment
- Heisenberg's Uncertainty Law
- The Time-Energy Uncertainty
- Quantum Super Position
- The Quantum Tunneling

Chapter 1: Introduction to Quantum Mechanics and Quantum Physics

Quantum mechanics or Quantum Physics is really an examination of vitality and significant issues at a fundamental level. A considerable law of quantum material science or quantum physics is that the vitality comes in composite bundles called quanta. Quanta works rather than the common marvel: particles can go about as waves, and waves keep showing up as particles.

Consequently, we can say that quantum mechanics or quantum physical science is the piece of material science that recognizes the littlest particles.

It carries with it all the manifestations of probably the strongest choices in the realm of Physics. With the size of particles and electrons, numerous conventional or Physical machines, which show how things move at ordinary speeds or sizes, stop to be significant. Quantum machines express those things as a rule that exist in some place and at a specific time. Notwithstanding, in quantum mechanics or quantum material science, things are fairly in the corner of chance; they may get an opportunity to be anyplace, to take a gander at you as source A, so one more opportunity to be elsewhere, suppose source B, and so forth.

1.1 Our Universe

In the mid-1900s and mid-1900s, the German logician Max Planck attempted to clarify the conveyance of shades that sparkled like never before in the light of extremely hot and white subjects, for instance, light filaments.

In understanding the circumstance in which he had chosen to introduce this spending plan, Planck recognized that it recommended that blends of specific shades (however a set number of them) were avoided, evidently those resulting from a complete number of a specific base. Some way or another, a few tints were estimated! This was surprising on the grounds that light was viewed as moving like a wave, which implies that the components of the haze must be a steady separation. What might permit particles to make tones between all these item numbers? This appeared to be strange until Planck believed quantization to be a simple science. As Helge Kragh brings up in her article, distributed in 2000, in the diary Physics World,

"Max Planck, the Reluctant Rebel"

In case of a disappointment with material science in December 1900, nobody came to see it. Planck was not absolved."

Planck's condition additionally contained a number that would seem, by all accounts, to be basic later on the advancement of Quantum Mechanics or quantum material science; today, it is known as "Planck's Constant."

Quantization has assisted with explaining the different riddles of material science. In 1907, Einstein utilized Planck's hypothesis of quantization to clarify why the vitality of an amazing individual changed at various rates if he set a similar temperature measure at what changed the first temperature.

Since the mid-1800s, spectroscopy research has demonstrated that various parts show up and produce more brilliant hues called "imperceptible lines." Although spectroscopy was a dependable strategy for deciding the segments of items, for instance, unavailable stars, scientists are befuddled concerning why all the segments light up those lines anyplace.

In the last part of the 1800s, Johannes Rydberg presented the marvel that indicated unfamiliar lines transmitted by hydrogen; notwithstanding, nobody can clarify why the circumstance works. This changed in the mid-1900s when Niels Bohr consolidated Planck's theoretical gauge into Ernest Rutherford's 1911 "planetary" model of the particle, which proposed that electrons spin around the environment similarly as the planets circling the sun. As per Physics 2000 (a site from the University of Colorado), Bohr proposed that electrons are caught in "surprising" hovers around the molecule center. They can "detonate" between various circles, and the vibrations made by the jump made a clear light, seen as alarming lines. Aside from the way that mathematical structures were viewed as futile science, they turned out to be quite certain and transformed into a norm for imagining Quantum Mechanics or quantum physical science.

What are Particles of Light?

In 1905, Albert Einstein conveyed the paper, "Concerning a Heuristic Point of View towards the Emission and Transformation of Light," wherein he imagined that light doesn't travel like a wave, yet is a type of "essentialness quanta." This mass of life, Einstein's suggestion, "can be imported or exceptionally created taking all things together," obviously when the molecule "bounces" between vibration levels. This will work similarly too, as will be demonstrated a couple of years afterward when the electron "bounces" between numerous circles.

Under this type, Einstein's "imperativeness quanta" contained the liveliness of bounce, when isolated by Planck's perpetual consistency, which energetic differentiation decided the light shade sent by those quanta.

With this brilliant perspective light, Einstein gave snippets of data on the continuation of nine novel wonders, including certain shades that Planck reflected in the arrival of straightforward fiber. It likewise explained how certain light beams could radiate electrons from metal parts, a wonder known as "photoelectric impact." However, Einstein was not totally urged to make this stride, said Stephen Klassen, an individual educator of material science at the University of Winnipeg. In the primary paper of the year 2000, "The Photoelectric Effect: Rehabilifying the Story for the Physics Classroom," Klassen states that Einstein quanta's imperativeness isn't fundamental to the acknowledgment of those nine miracles. Some light science medications, for example, waves are as yet being created to show the two hues Planck uncovers are removed from low fiber and photoelectric impact. To be sure, in Einstein's dubious triumph in the 1921 Nobel Prize, a gathering of Nobel laureates acknowledged his "photoelectric effect law," which didn't unequivocally rely upon the possibility of quanta.

Almost twenty years after Einstein's paper, the expression "photon" begat to communicate quanta life, on account of Arthur Compton's 1923 work, which indicated that the light sent by an electron obstruction changed from striking. This demonstrated the light particles (photons) needed to hit the emanation particles (electrons) in these lines affirming Einstein's expectation. Now, unpredictable light could have proceeded with both as waves and atoms, setting the light of "wave-particle duality" on the foundation of Quantum Mechanics or Quantum Physics.

What are Waves of Matter

Since the revelation of the electron in 1896, proof that all the components existed as particles steadily met up. Taking everything into account, the astounding showcase of light wave-atoms made analysts question whether the issue will undoubtedly work as particles. Possibly a wave-particle copy may sound powerful in this issue too? The best specialist who built up this hypothesis was a French researcher named Louis de Broglie. In 1924, de Broglie utilized Einstein's speculation for unordinary connections to show that particles could display wave-like properties and that waves could show atoms as images. Around then, in 1925, two scientists, working uninhibitedly and utilizing various lines of mathematical thinking, utilized Broglie's thinking to decide how electrons moved toward particles (a mystifying wonder utilizing old mechanical conditions). In Germany, physicist Werner Heisenberg (working with Max Born and Pascual Jordan) accomplished this by building "network hardware." Austrian physicist Erwin Schrödinger has built up a near idea called "mechanics wave." Schrödinger found in 1926 that the two strategies were comparative (yet Swiss researcher Wolfgang Pauli sent unpublished outcomes to Jordan, indicating that the framework gear was depleted).

The Heisenberg-Schrödinger molecule model, in which all the electron voyages like a wave (regularly alluded to as a "cloud") around the absolute particle that replaces the Rutherford-Bohr model. One particularity of the new model is that the end of the electron wave must be consolidated. In "Quantum Mechanics in Chemistry, Ed the Third." (W.A. Benjamin, 1981), Melvin Hanna states, "The weight of constraints restricts the energy in various characteristics."

The consequence of this clarification is that all pinnacles and rhinos are permitted alone, which clarifies why fewer structures are estimated.

In the Heisenberg-Schrödinger molecule model, electrons comply with the "turn capacity" and hold "orbitals" rather than the circles. Dissimilar to the Rutherford-Bohr model's round circles, do atomic orbitals have an assortment of situations extending from circles to free assets to daisy?

In 1927, Walter Heitler and Fritz London reclassified wave mechanics to show how orbital orbitals can participate in the arrangement of sub-nuclear orbitals, which unmistakably show why particle collaborate together to shape particles. This was another issue that couldn't be settled utilizing number juggling. These bits of information are committed to progress in the field of "quantum science."

1.2 You call it Magic; I Physics

The last point regularly prompts this: as unusual as it might appear, the study of quantum realism or quantum material science is, to a great extent, established that it isn't some sort of Magic. The things you anticipate are weird about the standards of general material science; however, they are altogether obliged by notable neighborhood headings and norms.

Thusly, there are different opportunities for an individual to go to the individual with the possibility of a "quantum" that appears to be ludicrous - free vitality, the capacity to fix witches, unfathomable driving of room - is substantially more likely than not. That doesn't mean we can't utilize material science to do astounding things - you can discover a truly cool new science as a rule - nonetheless, those things fit well inside the constraints of thermodynamics and simply the essential presence of the brain.

So you have the fundamentals of the quantum material science community. I may have overlooked a couple of things, or given a couple of discourses explicitly to fulfill everybody; in any case, this should be that as it may fill them in as a valuable first section for additional conversation.

1.3 The Uncertainty Principle

Further, in 1927, Heisenberg made another amazing pledge to material science or quantum material science. He contemplated that as the issue moves like waves, a couple of structures, such as the position and speed of an electron, are "viable," implying a breaking point (emphatically recognized by Planck) on how the precision of everything resources can be known. Under what may have been viewed as ordinary "Heisenberg's weakness," it was believed that by knowing the electron's position all the more precisely, its speed could be exactly decided, and something else. This danger law applies to typical size objects, yet isn't evident in light of the fact that the nonattendance of truth is shockingly little. As called attention to by Dave Slaven of Morningside College (Sioux City, IA), when baseball speeds are known to be inside 0.1 mph, the most noteworthy precision can be perceived when the ball position is 0.00000000000000000000000008.

1.4 Three Revolutionary Theories

Quantum mechanics or quantum material science have been created throughout different decades, all of which started as various questionable logical conditions and various points of interest of other numerical tests that couldn't be exact.

Everything started during the twentieth century, about a similar time as Mr. Albert Einstein distributed his vision and idea of relativity, which was an unmistakable change in the field of material science that normal the development of different articles at extremely high speeds.

As opposed to connection, nonetheless, the underlying foundations of quantum material science or quantum mechanics can't be credited to a solitary scientist. We can say that, maybe, numerous analysts have extended their work to set up three fundamental principles, until this point in time, that is worthy. These laws have instituted someplace in the period 1900 to 1930.

Description of Multiple Structures

A few properties, for instance, position, speed, and obscuring, may happen to a great extent just by irregular, fixed numbers, for example, dialing "click" from number to number. This has tried the basic idea of traditional hardware or old-style material science, expressing that such structures must exist smoothly and reliably. To show that it is workable for a couple of structures to be "clicked" as dials with clear settings, the scientists stated "quantized."

Light Particles

Light can now and again go about as a particle. This was initially met with an awful investigation, as it kept 200 years from getting research indicating that light demonstrations like a wave, like waves outside a tranquil lake. The light proceeds similarly by hopping up and contorting corners, and that pinnacles and waves can include or eliminate.

The stature of the introduced waves brings an excellent light, while the estimating waves produce peril. A light source maybe thought of as a soccer ball on a stick that is victimized by music with attention on the lake. The shrinkage delivered is identified with the contrast between the pinnacles, which is controlled by the ball's enthusiastic speed.

Waves of Matter

The story can likewise proceed with like a wave. This negates examinations of around 30 years that show that the issue (for instance, electrons) exists as a molecule.

1.5 Quantum Mechanics is Probabilistic

One of the most astonishing and (regularly, regardless) flawed parts of quantum material science or quantum material science is that it is hard to certainly anticipate the result of a solitary test in the quantum system. While researchers anticipate a specific test result, the forecast continues as before for all conceivable explicit outcomes. The connection between the theory and the nonstop test incorporates the spread of chances from numerous review tests.

The logical composition of the quantum outline regularly shows up as a "round work," frequently referenced with regards to the old Greek letter set PSI:

Ψ

We can say, there is a lot of conversation about what, decisively, this wave action you are discussing, separates the two primary camps: individuals who think about the wave work as a genuine physical article "psi-logicians") and individuals who think about the capacity of the wave as simply an assertion of comprehension (or absence of that office) corresponding to the basic idea of a specific quantum object ("dissident" theory).

In any basic model classification, the probability of acquiring the outcome isn't authoritatively given by the meandering capacity, yet by the square of the wandering capacity (uninhibitedly, by any measure; the wandering capacity is a logical confusion (for example, incorporates numbers that are not founded on a solitary negative square), and the likelihood work is incorporated additionally, however, the "square of the wave work" is sufficient to get an essential thought).

This got known as the "pregnant law" after the German rationalist Max Born initially suggested this (as indicated by a 1926 paper) and beat a couple of individuals as an uncommonly picked loathsomeness improvement.

There is a functioning exertion in specific pieces of the organization of quantum establishments to discover how to get the brought into the world norm from the principle direct; So far, none of this has worked consummately. However, it yields an intriguing science.

This is likewise essential for a legend that makes things like particles in various locales simultaneously. All that we can expect is a chance. Before the scale yields a specific outcome, the system is estimated in an unpredictable divulgence that coordinates all expectations by raising all expectations by numerous chances. It won't make any difference if you consider this to be a system in each region immediately. Being in one unclear spot is a lot of reliant on your emotions about ontic infection models.

1.6 Differences between Classical Mechanics and Quantum Mechanics

To sum things up, the fundamental contrast among quantum and old-style material science is the same as between an incline slope and stairs.

In old-style mechanics, occasions (in simple) are constant, or, in other words, they move in smooth, precise, and predictable examples. Shot movement is a genuine case of traditional mechanics. Or on the other hand, the hues of the rainbow, where frequencies progress consistently from red through violet. Occasions, at the end of the day, continue steadily up an incline.

In quantum mechanics, occasions (specifically) are erratic, or, in other words, "hops" happen that include apparently arbitrary advances between states: henceforth the expression "quantum jumps." Additionally, a quantum jump is a win or bust suggestion, similar to hopping from the top of one structure onto one another. Moreover, you either make it or you break it! Occasions in the quantum world, as it were, bounce starting with one step, then onto the next, and are apparently spasmodic.

For instance, electrons progress between vitality levels in a molecule by taking quantum jumps starting with one level then onto the next. This is found in the emanation spectra, where different hues, demonstrative of vitality level advances made by electrons, are isolated by dull zones. The dull territories speak to the region through which electrons make quantum - and along these lines, dis-persistent - jumps between vitality levels.

There are numerous different contrasts among quantum and old-style mechanics, including, for instance, clarifications of the alleged "bright calamity." Yet, these are too specialized even to consider discussing in detail here.

Let me simply state the last contrast among old style and quantum mechanics is the quantum thought of the "correlative nature of light," which expresses that light is BOTH a molecule, which has mass and a wave, which has none. This apparently opposing idea shows how weird quantum material science can be when contrasted with traditional physical science.

Chapter 2: Building Blocks of our Universe; Particles, to Waves, to Interference, and Fields

The expression "quantum mechanics" shows up in a 1924 paper by MAX BORN "Zur Quantenmechanik." Outside of discourse talk, it recommends that particles are machine contraptions. From the perspective of quanta and Quantum-Geometry modules, this articulation obtains the pith of particles.

Quanta modules are restricted estimation devices. The whole particle is described by a special instrument that is comprised of the consistency of quanta modules.

Fundamental proton articulation: The adaptability of quanta modules empowers the plan of different shells that address the particle's resting and compact atoms. Live streaming is made into creative, exchangeable, with contraptions sorted out by quanta modules. Broadcasting is additionally accomplished by the inclusion of voids and obvious parts. A correspondence system was created to survey possible possibilities inside sub-proton sub-segments. The framework shows that the joined tetrahedron/octahedron can't be sufficiently required to be depicted as a quantum-mechanical machine.

There are similitudes among particles and standard machines. A proton has a rotor part like a stator. The shut Octet analyzes to the Rotor, and the incorporated tetrahedron/octahedron resembles a stator. A proton can be considered as a modifier of a close-by void capacity. The proton conceals portions of this compound.

Hadron's degree as spoken to in "Quanta modules and material science" (1987). When considering wave atom, Earth particles or particles are prohibited.

As far as the proton, living space and inside particles are equivalent. With the restricted appearance time, kaon, pion, and muon, the settlement are nearby, while the polyhedra above allude to particles' mass. Inside the arrangement, these groups play oscillatory developments.

2.1 Fundamental of our Universe; "Particles"

In 1905, Einstein distributed a paper named 'Concerning the Heuristic Point of View towards the Emission and Transformation of Light,' where he figured light didn't travel like waves, yet as a specific 'amount quantum.' Einstein proposed that this boost could be 'presented or made more normal,' as the particle 'bounces' between the vibrated vibration levels. This will work a similar way, as allowed a couple of years later when the electron "bounces" between the tried circles. Underneath this model, Einstein's lively power fuses the differentiation of ricochet imperativeness; when isolated by Planck's soundness, the quality contrast decided the light shade sent by that number.

With this ideal approach to look at the light, Einstein gave insights into the behavior of nine different phenomena, including the specific colors on the best way to eliminate electrons from metal surfaces, a wonder known as the "photoelectric impact." However, Einstein was not given a lot of insurance against this heightening, said Stephen Klassen, a material science partner at the University of Winnipeg.

In a 2008 paper, "The Photoelectric Effect: Reconstruction of the Physics Classroom Tale," Klassen brings up that Einstein quanta's imperativeness doesn't clarify any of these nine miracles.

For example, some light-producing medications are prepared to uncover certain hues that appeared by Planck, for example, light-radiating fiber and photoelectric impacts.

To be sure, for Einstein's notable 1921 Nobel laureate, the Nobel Committee as of late observed "his disclosure of the law of photoelectric effect," which didn't rely altogether upon the possibility of dynamic vitality.

Almost twenty years after Einstein's paper, the expression "photon" was authored to communicate quantum size based on Arthur Compton's 1923 work. He brought up that light scattered through an electron section changed the diminishing. This indicated splendid particles (photons) crashing into electron particles and affirmed Einstein's reasoning. It was clear right now that light could travel like waves and atoms, setting a "wave-particle light" of light at the commencement of QP.

Rising Waves and Stationary Waves

Waves, for instance, discuss supposed 'moving waves' since they 'travel' in space. The model appeared, the improvement from left to right; nonetheless, it might be from left to right.

Like the influxes of the ocean, we should think about the 'standing waves.' We see that the wave has a similar shape as recently examined, and the water is free once more; however, the wave doesn't move, yet stays in a practically identical position - subsequently, the name. As a rule, a stop wave happens when it is obstructed by a 'gap' associated with two cutting focuses. A wave being built up shows up at one of the cutting focuses and is withdrawn toward another path. When the two-way waves are associated, the consequence of the net is a fixed wave. If all else fails, the open territory's dividers with the fundamental expectation that the wave can't assault them, and this makes the wave fulfillment be equivalent to zero at the limits of the hole.

1 This recommends halting flooding waves just in the drop - then again, actually, all together for the wave to be as low as could be expected under the circumstances, its repeat ought to be at tallness reasonable for the absolute number of pinnacles or posts to enter the space.

This law bolsters the improvement of different apparatuses. For instance, a note communicated by a violin or guitar is guided by a frequency gave by a wire, for which it is attached to the length of the strings the player puts on the impact. To change the notes' tallness, the player presses the string down to the next sharp focuses that change the length of a specific vibration period of the string. 2 Standing waves anticipate a similar capacity in each instrument: wood and metal breezes set vertical waves with moderate air volumes. At the same time, the drums' sound begins from the vertical waves set on the skin of the drum. The sorts of sound communicated by different instruments have passed and shifted - despite the fact that the notes made share something practically speaking. In view of this, we propose that vibration is in no way, shape, or form a reasonable 'indistinct' note contrasted with one of the permitted frequencies, yet is comprised of a mix of fixed waves, the total of its waves bringing about a sharp drop or 'head' rehash.

Stale waves happen when the wave is restricted to space. Now and again, it has become, however much as could be expected, not in space.

Regardless, if the waves were as yet the entire thing, the sound would not go to our ears. All together for the sound to be sent to the group, the instrument's vibration must make the movement waves obvious around it, communicating sound to the group. All around, for instance, the metal body vibrates with the affectability of the rope and makes a movement wave that associates the gathering.

A striking bit of science (or vitality) of altering instruments includes guaranteeing that the notes of the notes compacted by the waves consider static waves to emulate the relating travel waves. Full comprehension of the instruments and how they send sound to the group is an extraordinary point in itself, which we don't have to push ahead here. Famous clients ought to inform the book on science concerning music.

Influxes of Light

Different encounters incorporate tremendous electric waves, which are reflected by radio waves sending signs to our radios and TVs and light. These waves have various frequencies and frequencies: for instance, standard FM radio signs have a recurrence of 3 meters, while the light power sparkles at their range, being around 4×10- - 8 m of blue light and 7×10- - 8 m red light; different tones have frequencies between these components.

Light waves are not equivalent to water waves and sound waves on the grounds that there isn't anything against the vibrating mode (e.g., water, link, or air) in the models referenced before. There is no uncertainty that light waves are ideal for void space, as is clear from the manner in which we see the light from the Sun and the stars. This little wave material knows a basic issue with experts in the eighteenth and nineteenth hundreds of years. Some find that space is normally not filled, yet is supplanted by a concealed item known as an 'aether' that was an idea to help impact the light waves. In any case, this speculation started to stress when it was accepted that structures need to give similarly high frequencies in light that can't be implemented by the manner in which aether doesn't give protection to the improvement of articles (e.g., Earth in its drift).

It was James Clerk Maxwell, and in the 1860's he indicated that the hypothesis was silly. By then, the study of solidarity and interest was rehearsed, and Maxwell had the choice to show that it was completely contained in numerous unique situations (presently known as 'Maxwell's Conditions').

He likewise brought up that a solitary kind of reaction to these conditions breaks down the vicinity of the waves to the choppiness of electric and alluring spaces that can discover void space without the requirement for a facilitator.

The speed at which these 'electric' waves travel is constrained by the principle seasons of power and energy, and once this speed is settled, it was viewed as immense from the speed of light force. This has dependably improved the likelihood that light is electric waves, and it is currently justifiable that this model works similarly at different marvels, including radio waves, infrared radiation (warmth), and X-columns.

Matter Waves

The way light, regularly alluded to as waves, has sub-atomic properties makes French scholar Louis de Broglie estimate that the different components we, specifically, consider to be particles of wave components. With these lines, a brilliant light, frequently thought of as a surge of little particles like a slug, now and again, would travel like a wave. This misshaped see was first affirmed during the 1920s by Davidson and Germer: they passed the electron bar with a graphite gemstone and took a gander at an obstruction framework that was near the key level that was sent when the light was cut.

As we have seen, this material is significant in guaranteeing that light is a wave, so this test is a shred of speedy evidence that this model can be applied consistently to electrons. Afterward, close revelations were made in the properties of weighty molecule surfaces, for instance, neutrons, and it has now been set up that wave-atom holding is a typical material in a wide scope of particles. Without a doubt, even the most widely recognized articles, for instance, sand, soccer, or cars have wave qualities, in spite of the fact that in these cases, the waves are not completely accessible - predominantly on the grounds that the amazing reiteration has neither rhyme nor reason, yet in addition, as the exemplary style is made of particles, all with their own frequencies and furthermore every one of these waves is dependably cut and created.

We have seen that on account of light, the recurrence of wave vibrations is plainly identified with quantum power. Due to the yield waves, the redundancy emerges more diligently to deliver and harder to gauge. Or on the other hand, maybe there is a connection between wave reiteration and article power, with the fundamental motivation behind which the atomic power is higher, shorter the recurrence of the issue.

In the old waves, there is a perpetual stream of 'meandering.' Thus, in waves, the outside of the water closes, in sound waves, climatic weight impacts, and in electric waves, electric and alluring fields change. What is the inseparable sum because of the waves transmitted? The conventional reaction to this solicitation is that no measure of the body thinks about this. We can discover the wave utilizing theory and the condition of the physical sciences, and we can utilize our outcomes to foresee expected money related evaluations, anyway; we can't envision the wave itself, so there is no compelling reason to characterize it and ought not to endeavor to do as such. To broaden this, we utilize the term 'wave work' and not.

2.2 Waves

Quantum fields are quantum-theoretical theories of exemplary style fields. These are the two customary parts of Einstein's field practice and Maxwell's electric field. One more approach to make a gander at the quantization cycle is to at first overhaul field conditions (which are still customarily) corresponding to mathematical administrators who consolidate those numerical monetary standards (this edge did not depend on measurements/insights, no material science yet); nonetheless, when we 'manage' with the /accompanying noteworthy states of the controller, incorporating plans not found in the old style, we say something (endorsed by assessment) that these new, "ludicrous" (common, not numbered Vision) courses of action that reflect Nature, including all quantum see that repudiate customary idea.

There is a huge amount of establishments for the utilization of quantum field speculation. In any case, a typical hypothesis of traditional style convictions, which is one of our best (non-esteem) things for Nature contemplations. Second, the quantum field theory can speak to (perceptions, adjusted suspicions) the creation and crumbling of particles, non-logical proportions of a quantum material. Third, the quantum field theory is relativistic inherently, and "mysteriously" (less, simply rich measurements) manages complex issues that plague even hypothesis of the quantativistic atom.

Be that as it may, no, quantum fields are not viable with having any kind of effect. Quantum fields are significant. In the quantum field speculation, what we see as particles is basically a fascinating field of the quantum field.

Quantum electromagnetism is an unmistakable "convenient" theory of the quantum field. There are two fields in it: the electromagnetic field and the electron field. The two components meet normally, vitality and vitality are invigorated, and vitality is made or killed. Along these lines, for instance, what we normally observe as an electron-focused electron is a sure association in quantum electrodynamics between an electromagnetic field and an electron field, where the electromagnetic field loses quantum incitement, and the electron field assimilates its vitality, force, and yield power.

How might you catch the idea of the quantum wave of the issue?

What is a Wave?

- It typically quantifies the span of a sine wave, i.e., the distinction between the broke down waveforms.

- Frequency gauges how frequently the sine wave is reestablished in a second.

- Quantity quantifies the size of the frequency scale over zero levels.

- The stage decides the situation of the point on the wave in the second situation in the space, in the recurrence units.

How would I measure the Wavelength?

- Size A

- Wavelength λ

- Category Shift $\Delta\varphi$

Interruption

It is extremely useful to utilize wave impedance to discover superfluous allotments. If two wave surfaces are suspended, their non-wave pinnacles may abbreviate (gainful impedance) while confronting a higher worth, and the vessel, by and large, will radiate a wave (harming the snag). The example of ruinous and ensuing impedance in space makes it simple to envision recurrence.

It is an element of material science that the lifting power isn't identified with the thick gatherings of particles and the frail set in a solitary atom in a machine at some random time. The capacity of the wave, despite everything, demonstrates a genuine quantum object. This is one motivation behind why, sometimes, it implies that 'every cell isolates.'

Quantum theory can just recognize the likelihood of a specific result. Which of these expectations is at last expected in the essential inclusion of the overall political race and the regions. Just a couple of appraisals under similar conditions show cautious dissemination of chances, which is additionally taken out from Schrödinger's framework.

$i\hbar \; \partial/\partial t \; \psi(r,t) = (-\hbar 2/2m \; \Delta + V \; (r,t)\psi(r,t)$

All dissects to date have demonstrated on apprentices' level: square modulus $| \; \psi \; | \; 2$ of the state work ψ alludes to the likelihood of getting a quantum object during t space position and all the various boundaries contained in ψ.

Slender Film Disruption

The obvious impacts of impedance are not restricted to the twofold edged computations utilized by Thomas Young. The impact of a little film block is brought about by the light demonstrating the two zones isolated by a range equivalent to its size.

The "film" in a space can be water, air, or some other indistinguishable or strong fluid. In splendid light, the obvious impedance impacts are restricted to films with the extent of a couple of micrometers. The noticeable model is an air pocket cleaner film. The light reflected from the air pocket is a two-wave lift - one obvious on the front surface, and the other pondered the back. Two waves show spread and interruption into space. The size of the cleaning film decides if these two waves can meddle with help or in a dangerous manner. The full test shows that considering the recurrence alone λ, there is a helpful impedance of film thickness equals to $\lambda/4$, $3\lambda/4$, $5\lambda/4$, and destructive interference for thickness $3\lambda/2$.

As the white light enlightens the cleaning film, the concealing gatherings are viewed as different frequencies that movement through destructive hindrances and are isolated from the show. The mirrored light consistently shows as a comparing shade of the radiated recurrence (e.g., when a red light is produced with a damaging impedance, the splendid light shows up as a cyan). Thin-oil films produce a near impact on water. In Nature, the quills of winged creatures, including peacocks and fowls, just as the shells of specific creepy crawlies mirror light as the shade of the significant changes with a state of modification.

This is achieved by the restraint of intelligent light waves from misleadingly planned structures or by the wide assortment of show posts. Subsequently, abalone pearls and shells sparkle from the restriction brought by presentations from different pieces of the nacre. Stone stones, for instance, opal, show the flickering impacts of gleaming that originate from dissipating light from the commonplace instances of round particles.

There are numerous employments of mechanical impacts of light hindering impacts. The principal foe of inclusion is the focal point of the camera's center focuses on little estimated films, and recovery records taken to make the impedance of a risky showcase of clear light. Constant progressed inclusion, which incorporates different slight film layers, is made to transmit light only within a narrow range of wavelengths and thus fill as recurrence channels. Multilayer textures are additionally used to upgrade the presence of the mirror on infinite telescopes and laser optical gaps. Genuine interferometry methodology measures little changes in related isolation by taking a gander at turning shifts in light-hindering plans. For instance, the state of the Earth in the obvious parts is reflected in the segments of optical waves utilizing interferometry methods.

2.3 Wave-fields and Interference

Like the entirety of Nature, it is conceivable that comprehension has developed dynamically, and we think it is over for us people - nonetheless, we are not a gathering with a decent comical inclination. But, this can be extraordinarily diminished: it is restricted to living creatures with neurons. Few, except for the individuals who have been intensely moved to the wool polytheist, can guarantee that cell life alone is conceivable, or that plants have it (and I likewise incorporate reptiles).

This intercession after the beginning of Grand Climacteric since the nerves presented.

Regardless, we are powerless as far as where we are gaining little ground, which is 1.5 billion after the primary acknowledgment of Grim Climacteric; neurophysiology, with all the neuroscience in question, won't uncover anything thusly.

Other than that, there is a favorable position in our stockpile that we, regardless of all, have not yet utilized: the information loupe. Likewise, it has desires. At its most honed point, a wonderful spot in the region was found a few million years back by the Grim Climacteric sign. We will see, where we will see, where Evolution switched gears in its neuronal endeavors and began another cycle, exploiting a quirk in the yearly structure of the nuclear world.

Cognizance Polychrome

Since the mid-1960s, tasks have been progressing at NASA and at different school perception communities for the proof of shrewd life in space - the exercises of SETI. This is a purposeful exertion to channel the air into producing power with non-static, non-static plans, drawings with high false markings, which are evident to the logical marvels we envision. Given that these endeavors are paying off, it is a serious deal in the event that we can impede our messages. One such message, most likely, will exist - where life is on our farthest planet, and where it is. Besides, that is something we can manage; we may even search for a more mediocre clarification. All things considered, dear, in the matter of the idea of thankfulness, the main branch throughout everyday life - or some may state, its motivation - the message would be clear. We won't have the option to offer anything to detract from after the clarification. The ideal situation can give a showing of enthusiastic association, and is related to neurophysiological neuro-neurotic, and not without the help of many fitting suppliers.

Notwithstanding, we, all, have a characteristic feeling of what mindfulness is. It is a propensity to perceive the rest of the world and ourselves, an inclination to have a limitless measure of crafty - the green shadows of a glade, the smell of honeysuckle, and the proposal of a solid breeze. The pattern, including the acknowledgment of things - a flawless face, a flood of a winking hand, presently! The sound of any quiet voice. Besides, the propensity, including the consideration and coherence of time, our bliss, and the concealment of marvels and our will - the entire polychrome I.

What portion of material science can add to that expansion?

2.4 Quantum Physics and Life on Planet Earth

Quantum material science is likely the best scholastic accomplishment throughout the entire existence of human advancement; however, it appears to be an excessive number of individuals that it is very blocked off and the idea of annihilation. This is essentially an independent bend on pop-science researchers and writers. Now! When we take a gander at the study of a quantum material, we ourselves are underscoring odd and uncommon miracles: Schrödinger's catlike at the tallness of the "living" and "dead," Einstein's protection from God playing dice, a noteworthy association with the separation associated by the quantum catch. These things are amazing in light of the fact that they are normal; nonetheless, looking at them in the lab requires the division of explicit quantum outlines. It very well may be anything besides hard to perceive any association between these supernatural occurrences and typical regular daily existence.

In any case, the study of quantum realism is inescapable. The universe, as we discover it, works in quantum laws. While the prehistoric studies of the old-style while applying material science to quantum material, numerous particles appear to be totally changed, numerous obvious, basic marvels need to manage quantum impacts. Here are a couple of instances of things you may have involvement with your day by day existence without understanding that they are quantum:

Toasters

The red light to warm the part as you fry a little portion or bagel is a visual impression for a significant number of us. This is additionally where the study of realism started: to explain why hot articles shimmer that particular shade of red is an issue for material science to comprehend.

The warmth exchanger of a warmth exchanger is an instance of a sort of fundamental, far and wide wonder that takes into account think doctors: regardless of what the item is made of on the off chance that it can withstand the warmth of a specific temperature, the width of the light we send is equivalent to another. Those basic practices originated from the best researchers of the last part of the 1800s; however, none of them got an opportunity to take care of the issue.

The way that the light was autonomous of the piece recommended an overall clear methodology: It consolidates all the shades of light that can be created by an item and gives every one of them an equivalent worth of the heat energy contained in the object. The problem with this is that there are more few ways to emit high-frequency light that the light of low frequency, which suggests that instead of a pleasant warm red glow, your toaster should be spraying x-rays and gamma rays all over the kitchen. That doesn't occur (something to be appreciative of!), and something else needs to occur.

The response to this issue was found by Max Planck and introduced "quantum theory" (giving the last theory its name) that the light could only be emitted in discrete chunks of energy, integer muntries of short. With a high repeat light, this quantum of vitality is more conspicuous than the part of the glow allotted to that repeat, and therefore, no light is communicated to that repeat. This cuts the light of high recurrence and accordingly advances the figure that interfaces the spectrum of light from hot objects with great precision.

Every time you toast bread, you're looking and observing a place when quantum mechanics or quantum physics start taking place.

Splendid Lights

The good old spotlights produce light by getting a little wire sufficiently hot to enlighten with a brilliant white light, which is their lights as a quantum in a reasonable toaster. You get light from another striking quantum measure in the event that you have splendid lights around long chambers or present-day twisty CFLs.

Returning to the mid-1800s, researchers found that all items in the time table had a unique spectrum. In the concerned event that you need to accomplish something, you must have a ton of time, once in a while, for everything. You get a vapor of atoms hot, they emmit light at a smallish number of discrete frequencies, in one model for everything. Such "undisclosed" lines were quickly used to gather a bit of dark material, and even the closeness of dark items - for instance, helium, was at first observed as a before unknown spectral line in light from the Sun.

As incredible as this seemed to be, nobody could clarify it until 1913 when Niel Bohr concocted Planck's quantum thought (Einstein extended in 1905) and presented the fundamental quantum model of the molecule. Bohr brought up that there are numerous uncommon cases where an electron will catch nucleus of an atom and not be gotten again when it is in those nations. The recurrence of light sent or info relies upon whether it tends to be utilized to have a particular recurrence of a particular range of specific particles.

This has been a progressing idea, yet it has functioned admirably to show how the segment advancement and quantum science has proceeded. While the peripheral picture of what occurs inside an atom is totally extraordinary comparable to Bohr's fundamental idea, the focal idea is the equivalent: electrons travel between heavenly states inside particles by sending light to specific frequencies.

This is the principal thought behind fluorescent lighting: Inside a light (either a CFL or a long chamber), there is a dab of mercury fire that is fueled into plasma. Mercury originates from light that frequently falls on a noticeable separation in a manner that can enlighten our eyes from the catch. The light looks white. In situations where you take a light utilizing a restricted light, as it ought to be clear in the interest mirrors, you will see some shaded bulb pictures, where the light makes a consistent rainbow light.

In these lines, when you utilize splendid lights to enlighten your home or office, you have quantum logical material for which you can be thankful.

Chapter 3: Everything is Quantized!

3.1 Drop by Drop; The building of our Universe!

The supposed Planck extends, excessively short for anything I could discover. Similarly, as with the advances made in innovation to develop the "speculation of everything," space isn't a problematic progression. It isn't smooth yet granular, and Planck's length gives the size of its small letters.

The time it takes for a splendid light to enlighten over this unobtrusive region (around 10 seconds to 43 seconds) is known as Planck's time, the incomprehensibly limited tick of the reasoning clock. Join these two thoughts and the proposal that reality is shapely. What is frequently thought of as the vacant hole works from tiny or quanta units?

"We've generally imagined that space-time ought to be determined," said Dr. Steven B. Giddings, a teacher at the University of California at Santa Barbara. "Ceaseless improvement has made motivating new suggestions on the most proficient method to make these thoughts more sensible."

The hints of the grain diet originate from endeavors to join a more extensive blend, Einstein's hypothesis of gravity, and quantum innovation, which shows the working of three distinct powers: electromagnetism and the solid and frail nuclear bonds. The outcome will be an autonomous structure - frequently called quantum gravity - that characterizes all the particles and powers of the Universe.

What is generally dicey about these combination endeavors, the superstring theory and the lesser-realized strategy called circle quantum gravity, both unequivocally suggest that space-time have a shorter building.

Nonetheless, what the break is probably going to resemble is that researchers are focusing on their psyches.

It isn't unexpected, at that point, that oats are regularly neglected in the indigenous habitat. Surely, even the littlest quarks that makeup protons, neutrons, and different particles are too huge to even consider detecting expected thumps on the Planck scale. Notwithstanding, particularly since it is past the point of no return, researchers have recommended that quarks and all that else are made of little materials: vibrating ropes with ten measurements. At Planck's level, the weaving of the room time would plainly show when the fine Egyptian cotton showed up under the amplifying glass, uncovering the wind and woof.

Uark and lepton, development squares, are shockingly little. In reality, even the biggest quarks are just about an attometer (million meters) wide. Nonetheless, zoom in - a billion times - zeptometer and yoctometer past, where these parts are anonymous. By then, go further, triple it, lastly hit the base: This is Planck's stature, about 1.6 x 10-35 meters, acknowledged by researchers to be the most available tallness known to humankind. They have ventured to such an extreme as to state, the general concept of detachment is of no worth.

How little would we be able to state we are discussing? It would take more time for Planck to cross-grain of sand than to take grains of sand to cross the noticeable Universe.

In light of everything, it is conceivable that the cut-off size may appear to be odd. Taking everything into account, if conceivable, you could construct a split in the center - at long last, alright? Not generally. Another uncommon disclosure of the 20th century is that little spaces, numerous physical structures, for example, incredible quality and essentialness, can take on certain particular characteristics, or "quanta." That level — kept up by numerous long periods of examination — is the development of quantum mechanics.

This proposes an inside and out examination: If this issue's properties are not determined, ought not something be said about the surface of the space itself? Is the Universe a consistently smooth surface, as confirmed by Einstein's hypothesis of relativity? Or, on the other hand, furthermore, if we somehow happened to look truly close, could everything be an impression of a shining pixel like a PC screen? Is the truth we are just taking a gander at a dream that contains little spaces?

Investigating the Planck scale with a nuclear smasher can take a metal the size of our earth. Regardless, scientists at Fermilab, close to Chicago, have another, imperceptible Hollow-meter device that can precisely catch a couple of snippets of data. Utilizing strong state lasers and some certainly cleaned glasses, they might want to get the undeniable jitter of those hypothetical pixels - delegated "holographic commotion," after a more point by point look.

3.2 Double-Slit Experiment

A case of a brand name of dull and staggering edges is seen when a monochromatic light that surpasses the limited cut enlightens an out of reach screen.

This obstructing configuration is achieved by an overlay structure that covers the light waves exuding from these two cuts.

Circuits of valuable impedance, like the splendid edges, are shaped when the distinction in the way from two slices to the edges is the most significant number of light voyages.

Risky blockages and dull edges happen when the title opening is the number of essential frequencies.

The impression of impedance impacts unquestionably demonstrates the vicinity of the tides. Thomas Young suggested that light is a wave and relies upon the law of worship; his extraordinary trial accomplishment was to exhibit the helpful and risky impeding of light. In adjusting to Young's scientific cutting, which movements to his stray pieces just from the light source, the Laser also enlightens two equivalent cuts in the perilous zone. The light that produces two cuts is noticeable on the distant screen. When cutting 'width' is more perceptible than light recurrence, numerical light targets hold - light activities, two shades, and there are two enlightened areas on the screen. Then again, as the cuts are littler in width, the light moves to the shadow of the figures, and the floods of light meet at the edge. (Trouble itself is achieved by the idea of light force, which is one of the inhibitory impact states, which is analyzed in more detail underneath.)

The speculation of the superposition clarifies the accompanying case of intensity on the lit-up screen. Helpful impedance happens when the way of the way from the two direct bleeding edges surpasses the necessary number of frequencies $(0, \lambda, 2\lambda, \dots)$. The irregularity of these examinations implies that the two waves' highest points show up all the while. Risky impedance emerges from an examination of the technique equivalent to the number of basic frequencies. The adolescent utilized factual contentions to show that the two waves' quality achieved the amassing of similarly scattered gatherings or the edges of superpowers, identifying with useful incense spaces, isolated by thick locales of incredibly perilous disallowances.

The recurrence of light force λ in the dispersion of the cut d is the essential limit in the two cut figures. In the event that the λ/d is under 1, the differentiation between the ceaseless edges of impedance won't be excellent, and the impacts of the restraint may not be obvious young was able to separate the interference fringes utilizing a barely separated. Consequently, he decided the size of the shadows of the obvious sun. The short frequencies of noticeable light speak to why impedance impacts are seen distinctively in clear conditions - the contrast between the light wellsprings of hindering light should be especially little to recognize fine territories and incense issues.

Taking a gander at the impacts of impedance is attempting a result of two distinct issues. Practically all light sources transmit a steady measure of recurrence, getting various impedance-covering plans, each with an assortment of diffused components. Distinctive square plans wash out the most characterized impedance impacts, for instance, locales of absolute murkiness. Second, for a model of impedance to be found in a sweeping period, the two light sources must be reciprocal. This implies basic sources will keep up a steady stage relationship. For instance, the incorporation of two shows of a similar reiteration consistently has a phase relationship in each space, either in front of an audience, off stage, or in an ordinary relationship.

Regardless, most light sources don't transmit genuine waves; rather, they emanate influxes of anomalous waves that change a few times each second. Such light is called unwanted. Impedance normally happens when light waves from two interconnected sources meet in space, yet the example of blockage increments quickly as the period of the waves changes suddenly.

Light signals, including the eyes, can't distinguish the elements of the copies of the plans, and they can just glance at the intensity of a typical time. Laser light is generally monochromatic (containing a solitary recurrence) and is profoundly justifiable; consequently, it is a fitting spot to distinguish the impacts of anticipation.

After 1802, Young's technique for estimating the recurrence of perceptible light can be joined with a quick estimating affirmation of the speed of light being reached around then to recognize the distinctive thickness of light. For instance, the recurrence of green light is roughly 6 × 1014 Hz (or cycles/seconds). This duplication by different significant degrees is a lot more prominent than the recurrence of typical waves. Independently, individuals can distinguish sound waves with frequencies of up to 2 × 104 Hz. For the last 60 years or somewhere in the vicinity, what truly occurred at that significant level stayed a mystery.

3.3 Rutherford's Experiment

In 1911, Rutherford and his partners Hans Geiger and Ernest Marsden started a progression of eradications tests that could totally change the celebrated particle model. They assaulted little sheets of gold managed with moving alpha particles.

In light of the affirmed atomic model, when the size and charge of the atom were consistently disseminated all through the ota, the scientists expected that alpha particles would go through the gold chain with almost no shirking. Some were even redirected back to the well. No earlier data has accommodated this disclosure. In the aphorism, Rutherford proclaimed that "it resembled shooting a 15-inch [15 cm] shell with a bit of tissue, and it returned and hit you."

Rutherford was anticipating thinking about a totally new atomic model to explain his outcomes.

Since, far away, the majority of the alpha particles had gone through the gold, he thought most about the atom was vacant. Then again, the particles that were re-coordinated and firmly coordinated were not yet in contact with the solid braking power of the dirt inside the atom. He imagined that all the positive charge and size of the particle mass ought to be taken out from the littlest space inside the phone, which he called complete. The core is a little thick, molecule focused focus comprised of protons and neutrons.

Rutherford's atomic model is known as the nuclear model. In the nuclear atom, protons and neutrons, containing practically any molecule size, are found in the particle segment. Electrons are appropriated around the center and have a lot of molecule volume. It merits stressing how little the complete contrasts and the rest of the aspect of the atom. If we somehow managed to break a particle into the size of a huge football field, the center would be the size of a marble.

Rutherford's model, in the end, turns into a critical improvement in the full comprehension of the particle. Other than that, you haven't totally managed the possibility of electrons and how they have expended the immense space around the center. With this and an alternate encounter, Rutherford was granted the Nobel Prize in Chemistry in 1908.

3.4 Wave-Particle Double-Behavior

The study of realism tries to explain rules as it focuses on improvement and matter. Regardless, quantum material science is attempting to comprehend the structure of minuscule particles and how they move. Such particles contain components, for example, electrons, protons, and neutrons.

Quantum Physics; Minute' Details

In its accentuation on non-perpetual particles, the study of materials determines the particles that make up little particles. Rules managing outwardly debilitated structures have plainly been mistaken in deciding the area of low-lying zones since the start of the 20th century. "Quantum" starts with the Latin word for "esteem." Material science is utilized to allude to little units of yield, and the life of their work is normal and found in quantum physical science.

Altogether, even the most insistent and reformist circumstances, for instance, are practical, despite the fact that they appear to be little.

The particle quantum model is significantly more befuddling than what we have seen previously, instead of spinning around a star-like center, electrons, and a hover in undetectable, lesser-known, or cloud-like turn of events. Also, the last arrangement we got in the electron assortment (alluding to the number of external shell electrons) is frequently bound to be then to a fixed request.

This carries us to where we comprehend the quantum idea of light to mirror the logical name of the material, so you can comprehend that its thought process is to show the capability of the electron space at some irregular time. Along these lines, when the word is related to "light of light," you should have a solid comprehension of the overall guideline of law.

It lives alongside Quantum Physics

The likelihood that contemplating anything can impact the physical cycles that happen is not the same as the study of realism. For instance, in what is known as wave-atom duality, light waves go about as particles, and these particles moreover go about as waves. Put another way, light has the qualities of two particles and waves, and even the clearness can deliver the capacity of light.

In the quantum upgrade, the issue can move between various places without moving from space to both. This gives away to the current application where the information can be isolated by a different split. As per quantum science, we find that the tone of the Universe can be alluded to as a continuation of possibility.

There are numerous kinds of material science. The one that centers for the most part around the conduct of lights (photons) is known as Quantum Optics.

By researching Quantum Optics, you will find that instances of the improvement of individual photons (light bars) straightforwardly influence the agreeable light. An essential and adaptable instrument known as LASER is only one of the numerous critical symptoms in Quantum Optics.

This is as a distinct difference with the more broad light examination, Classical Optics, by Sir Isaac Newton, in which light seemed to have just sub-atomic structures, implying that it moved in a methodical manner, returned to objects, and went through items with immaterial obstructions.

Photons

To make it more clear what is done when the word photon is utilized, let us guide our focus toward the Photon Theory of Light. Photon is a canny (or quantum) parcel of electrical (or glowing) vitality in this specific sense.

Situated on a clear and stable machine, photons have a consistent light speed for all watchers. It happens at the speed of light (particularly, specifically, alluded to as the speed of light), the term by which it is utilized.

C = 2,998 x 108 m / s height

Altered in Photon Theory of Light, the principle attributes of photons are as per the following:

- They move at a consistent movement, c = 2.9979 x 108 m/s (low speed) in the free space.

- They are known to have zero trouble and eagerness.

- They communicate vitality and vitality, contrasted with the recurrence of nu and recurrence lambda, (and p, vitality) of an electromagnetic wave by

E = hv ye no p = h/lambda

- It might be destroyed or created when radiation as absorbed or emitted.

- They have the ability to have particle-like interactions, for instance, electron impacts and other transitory sections.

Quantum Optics; Basic Understanding

To all the more likely comprehend the quantum properties of light, it can assist with incorporating part of the relating cycles (maintenance, yield, and vivified yield) into the Laser, as this is one of the most surprising employments of quantum optics. For the most part, these three equivalent credits can be summarized in various light sources at expanding levels.

Electronic advances are normally a sort of progress that communicates or consolidates obvious light. Simply imagine an electron moving between levels of atomic size to perceive how this functions.

For Laser to work appropriately, vivified light yield is fundamental. Sustainable light yield is utilized to give the advancement expected to play out a basic reasoning capacity.

The tale property known as levelheadedness is the aftereffect of an amended leave rate. Ordinary advancement makes discharge times that are needed to give improved light. This fixes the delivered photons in a decent pre-request plan where all the photons have a full stage relationship to one another.

This sort of insight (related arranging) is described by two unmistakable terms: impermanent affectability and area mindfulness. Both wind up being critical in the advancement of the blockchain used to produce perceivability.

3.5 De Broglie Hypothesis; Time and Space! Are they Quantized!

This was one of the most acclaimed logical meetings ever. Of the 29 up-and-comers, 17 got or got Nobel prizes. The gathering is significant for two titans of material science: Niels Bohr and Albert Einstein.

1927 was per year, and researchers were stunned. The very presence of such an astounding thing is in peril. Are electrons, lights, and comparable articles, waves, or particles? In certain tests, the little bodies act like waves, and in others, they act like particles. This isn't going on in our large world. The sound waves don't act like rocks - and fortunately, your ears would nibble at the present time.

The 1927 Quantum Mechanics meeting talked about a blend of terms that appeared to be conflicting. Schrödinger and de Broglie introduced their perspectives. Be that as it may, 800 gorillas were Bohr. It was later called the Copenhagen interpretation. Bohr recommended that wave estimations were characterized as materials, for example, electrons; however, as particles, associations didn't exist until somebody needed them. The demonstration of appropriation turned into the beginning of life. Utilizing Bohr's own words, the individuals included had no "obvious life in the typical setting." None of that would have been Einstein.

Einstein would not have had that. The electron was an electron, and in light of the fact that somebody was not taking a gander at it, it was still there - any place it "was." Towards the finish of the meeting, Einstein tested Bohr's view. Yet, that was just the start. When thirty, Einstein was dead, Bohr and Einstein were entangled in warmed dealings - eye to eye and printing.

The discussions were of a courteous fellow. Bohr and Einstein were old buddies and regarded each other definitely. Notwithstanding, they persevered.

He stated, "It's not reasonable to believe that material science needs to discover what nature resembles," Bohr said. Einstein opposes this idea. "The main reason we disclose to science is to discover what it is."

For all its unpredictability, Bohr's meaning of Copenhagen stays one of the world's most broadly acknowledged quantum material science ideas. Numerous normal definitions seem like most outsiders. In any case, they all highlight one straightforward truth. Our Universe is a secret, as all researchers will let you know. It derides us with unfathomable realities and gives us meaning. Possibly one day, we'll go to it. However, we should confront the great secrets around us before that.

Moreover, Planck's time is the fundamental unit of time in the arranging of Planck Units. Significant:

$t_p = 5.39 \times 10\text{-}44$ s

In SI units, time estimations are made quickly (generally given s pictures). Aside from the way that the utilization of seconds has the benefit of everyday presence, for instance, estimating the time it takes for a contender to run 100 meters or the length of a telephone, notably, it is little on the planet when we talk about ordered occasions in the early Universe, for instance. Which occurred in the 10-35s after the Big Bang).

The consequence of utilizing seconds to quantify time is that huge changes take esteems that are not frequently accommodating in recalling circumstances:

Light speed c = 299792458 m (s) (s)

Gravity G = 6.673 (10) x 10-11 m3 kg (- 1) s (- 2)

Board Strength (diminished) ħ = h/2π = 1.054571596 (82) x 10-34 kg m2 s-1

Boltzmann strong k = 1.3806502 (24) x 10-23kg m2 s-2K-1

Planck's time is resolved to utilize the size of a mix of these key components:

By revising the base units long, size, and time comparative with Planck units, the main points of interest are:

c = G = ħ = k = 1

Presently, Planck-Time is the time it takes for a photon to have any kind of effect equivalent to the length of Planck:

= 1.62 × 10-35 m

This is the briefest time limit conceivable. With its overall length of Planck, Planck's time characterizes the scale at which the current assemblage of thought is clamoring. At this level, the absolute time figures are as wide as the relative associates. Along these lines, on such scales, to date, vague speculation that consolidates typical cooperations with quantum machines is relied upon to mirror the laws of material science.

Accordingly, our present introduction of the main Universe improvement starts at tp = 5.39 × 10-44 seconds after the Big Bang.

Chapter 4: The Heisenberg's Uncertainty Law

The standard of Uncertainty, otherwise called the Heisenberg Principle of Uncertainty or the Principle of Indeterminacy created by German thinker Werner Heisenberg (1927), states: or in principle, the shape and speed if issue can't be clarified all the while. The very meanings of the exact area and the specific extents where they meet up are unessential.

Conventional experience doesn't mirror this rule. The state of the vehicle and its position is agreeable in computation on the grounds that the vulnerability related to this perspective on everyday things is little to such an extent that you can see it. Complete law specifies that the result of exposure is equivalent to or more noteworthy in position and speed than the base or persistent worth (h/(4μ) in which Pl Plk constants or roughly 6.6/10−34 seconds). The impact of shakiness just applies to little fields of iotas and sub-nuclear particles.

Any endeavor to precisely figure the speed of a molecule under a particle, for example, an electron could unexpectedly affect it with the goal that its position would not be permitted to compete at the same time. These discoveries or inventions have nothing to do with the insufficiency of estimation, cycle, or survey devices; in light of close common contact of particles and waves with subatomic size.

Every molecule has a wave; all particles have a wave-like nature. The molecule is most regularly discovered when waves are enormous or substantial. Also, the more noteworthy the frequency, the more prominent the frequency gets more pronounced, and the molecule pressure is set up. Only a spotless wave is of unending length; despite the fact that its comparing molecule has a specific position, it has a particular speed.

Then again, molecule wave with all around characterized frequency engenders; a similar molecule can be anyplace, despite the fact that it has a specific speed. A precise estimation of a solitary difference alludes to the overall vulnerability while figuring different factors.

The idea of vulnerability is likewise communicated as far as elements and molecule particles. Molecule pressure is equivalent to the result of its size and speed. In this manner, the vulnerability's impact is comparable to or higher than $h/(4\alpha)$ in the force and molecule position. The guideline applies to other equal sets of perceptions, for example, quality and time: the result of vulnerability yet to be determined of intensity and exposure over time of examination more prominent or equivalent to $h/(4\beta)$. On account of a precarious particle, a close association happens between the Uncertainty of the measure of radiation and the danger of the unsteady framework as it prompts stable advancement.

Chapter 5: Quantum Super-Positioning

At whatever point you play the guitar and hear the agreement, you experience the waves' impacts. The hints of every arrangement consolidate as they arrive at your ear. On the outside of the lake, in the wake of tossing a little stone, something very similar occurs: the knobs meet and meet on their shoreline.

Sound waves and water waves are raised, complete by singular wave focuses framing another wave.

These two conditions have one shared factor: the rotating waves consolidate to coagulate their plummet. The outcome is a point-by-point proportion that delivers another wave.

Waves can depict molecules, electrons, and a few different occupants of the quantum universe. Yet, these waves don't demonstrate the development of people like water or wind. Instead, their mobile pinnacles and valleys may have total qualities estimated by quantum resources, for example, position or force. The electron iota is showered into the orbital haze of chance.

For instance, the electrons circling a particle don't exist anyplace known to man as do the Earth when it orbits the sun. Preferably, it is set in an orbital haze of chance. This space cloud is a practical 3D quantum wave, comprising of mountains and valleys that change after some time and speak to the opportunity to get electrons in a given space.

The calculation of this wave shifts as per the quality of the electron. A surface can be made where two quantum waves - speaking to two degrees of electron vitality - are assembled, prompting another example of pinnacles and valleys. This changes where the electron is well on the way to be found and the noticeable structures of the particle can be influenced.

Steve Rolston, President of the Department of Environment at the University of Maryland, clarifies why our everyday experience doesn't have a quantum scale.

It isn't unexpected to state that maybe an electron has two unique energies simultaneously or that it is in a few places simultaneously in this kind of amplification. In the event that you consider electron only a molecule, this won't be clear. In any case, when you think about an electron as an all-encompassing item, the overlay is more straightforward. Waves - including wave super-positions - are in numerous spots simultaneously.

The setting may appear to be oddly extraordinary now and again, for example, putting an apple close to orange and attempting to call a banana; however, it is valid.

This is not, at this point, clear in standard quantum tests. Along these lines, a separate electron shaft (or another quantum molecule) is terminated into a layer containing two little sequential cones — the delicate identifier records when an electron hits the opposite side of the cuts.

At the point when electrons act like particles - consider little ball balls - so you can hope to see an example of two arrangements of locators with a set behind each space. At that point, the identifier follows the unsettling influence, as though every electron ventures like a wave between the two limits.

The wave goes through the two bumps simultaneously. The resulting aggravation makes numerous dull and splendid regions on the divider.

This is a test that requests feeling; however, quantum material science shows what occurs. These two parts express the substance like the conduct of every person by driving every one of them into a voice over the flap "past the left projection" and a "right lumbar" space ("right projection").

A Quantum object goes about as a wave and molecule. Slides are focused on singular particles, yet the following example is like that of a wave.

Chapter 6: Time-Energy Uncertainty

Another type of Uncertainty theory is related to uncertainty when simultaneously measuring the strength and health of a quantum condition.

However, the general significance of the theory of the time of power is that the emerging quantum state is one of many times. The aim is that the state's frequency is equal in proportion to the time and that the frequency is connected to the state power and that the state must, therefore, be considered in many cycles to measure power accurately.

Think of the pleasant atmosphere you can display. The limited lives of these provinces can be determined from the configuration of radiation boundary lines. - When the excitement level decreases, the emission energy varies slightly, so the emission line is reflected by the spectral (or wavelengths) emitted photons emitted. Displays all spectral lines by spectral range. The photon's total energy corresponds to the potential energy of the excitement mode and provides direction to the main extraction line. The short-lived regions have a more comprehensive spectrum range and a wider range of long-term areas.

Chapter 7: Quantum Tunneling

Tunneling is a function of mechanical quantity. A tunneling current occurs when the electrons in it move across a barrier that they cannot move classically. If you have no energy to "overcome" an obstacle in classical words, you will not. Anyhow, in the quantum mechanical world, electrons have wave-like properties. Such waves do not stop suddenly, but gradually taper off at a wall or a barrier. If the barrier is relatively thin, the probability function will enter the next area via the barrier! Because an electron is low on the other side of the barrier, other electrons are traveling and emerging on the other side. It is called tunneling when an electron passes through the barrier in this way.

Quantum physics tells us that electrons have wave-like properties and particle-like characteristics. Tunneling is a wave-like result.

When an electron (the wave) hits a boundary, it does not stop suddenly, but it tapes rapidly and exponentially. If the barrier is thick, the wave won't pass. Part of the wave passes, and some electrons may occur on the other side of the barrier.

The number of electrons going to pass through the tunnel depends on the width of the tube in which it is loaded to the body. The sum of electrons that are pumped into the tunnel depends heavily on the size of it in the pipe in the tunnel. The present continuously slides through the fence, with the barrier's width.

To extend this full description to the STM: the main starting point of the electron is either the sample or the tip, which depends on the instrument's setup.

The boundary is the distance (vacuum, air, liquid), depending on the experimental setup, the second area is the reverse, i.e., sample or edge.

The size of the current is measured by measuring the current through the distance.

Piezo-Electric Effect

In 1880 this effect was discovered by Pierre Curie. The result is the crushing of the edges of individual crystals such as quartz or titanium barium. The effect is that opposite charges are generated on the sides. The outcome may also be reversed; the strain applied to a piezoelectric crystal is lengthened or compressed.

Such components are used to scan the tips in microscopic scanning tunneling (STM) and other scanning research techniques. PZT (lead zirconium titanate) is a typical piezoelectric substance used in the scanning method's microscopy.

Chapter 8: The Black Body Radiation

Warmed Bodies Radiate

For the present, we will go to another riddle that disdains researchers as the new century turns (1900): how warm bodies start? There was a finished comprehension of the framework being referred to - the heat was known to make particles and particles vibrate enthusiastically, and particles and atoms were demonstrated instances of electrical charges. (Clearly, Newton was fit as a fiddle.) From the examination of Hertz and others, Maxwell's possibilities for light-emanating redirection cases have been affirmed, regardless. It was known from Maxwell's conditions that the radiation went at the speed of light, and for this situation, it was perceived that the light itself, alongside the warm beams related to the field, was really electric waves. At that point, the image was that when the body was warmed, the resulting vibrations on the sub-nuclear and atomic-scale unavoidably eliminated—recognizing at the time that Maxwell's concept of electromagnetism, which is the most effective in the physical world, was genuine. At the sub-nuclear level, these attractive expenses would have passed, maybe radiating warmth and noticeable light.

How Is Radiation Absorbed?

What is implied by the articulation "dull body"? The truth is that the hot body's radiation relies here and thereupon upon the body being warmed. We ought to quickly uphold and consider how various materials store radiation to see this with incredible achievement. For example, a couple appears to get light in any capacity - light passes straightforwardly. With a sparkling metal surface, light is excluded.

It might be noticeable. Dim materials, for example, debris, light, and warmth, are totally packed, and the hardware is warm.

How might we comprehend these different cycles, for example, light waves that adjust to changes in applications, making these charges influence and store vitality from radiation? On account of the glass, unmistakably, this isn't going on, in one way or another, practically nothing. Why not? Full comprehension of why it requires quantum gear; notwithstanding, the overall thought is as per the following: there are costs - electrons - in the glass that can change in the light of the influenced electric field outside, yet these charges are immovably appended to the particles, and can just vary in specific waves. (In quantum craftsmen, these charging vehicles happen when the electron moves starting with one circle then onto the next. Recognizable, so there is no repeat with a little wave, and starting now and into the foreseeable future, the energy is gotten. That is the reason glass is ideal for windows! Duh. Notwithstanding, the glass isn't sure about specific waves outside the noticeable separation (generally speaking, both infrared and light). These are waves where the conveyance of power charges on particles or bonds can frequently vacillate.

How might we comprehend the considering light through metal? A little metal has electrons permitted to go at all forces. This is the thing that makes iron into metal: it conducts vitality and warmth viably; the progression of these straightforward electrons really sends both. (In light of everything, a little warmth is moved to the vibrations.)

But metals are overwhelming on the grounds that they shimmer - why would that be? Indeed, it is those free electrons: they are squeezed into bigger portions (contrasted with particles) by the electric field of the moving toward light waves, and this animating stream originates from the electric field, much like the stream in a talking radio wire. The radiation is the mirrored light. With a sparkling metal surface, a slight shine of fabulousness is joined with the warmth, it is consequently recharged, and that is it is obvious.

Right now, what might be said about taking a gander at something that focuses light: there is no transmission and no showcase. We are moving toward the best end with cinders.

Like steel, it will lead to the progression of power, be that as it may, not so much as an effective methodology. There are detached electrons, which can go at all energies, yet keep on holding objects - they have a momentary significance. At the point when they thumped, they caused an upheaval, similar to the balls hitting the watchmen on a pinball machine, so they emitted a solid power in the warmth. Aside from the way that the electrons in the debris have a shorter length contrasted with that of the honorable metal, they are plainly contrasted with the electrons limited by particles (as in glass), so they can quicken and pick up vitality in the electric field. Along these lines, they are ground-breaking go-betweens in moving vitality from light waves to warm.

8.1 Absorption and Emission

Subsequent to perceiving how the remains can get into the beams and move vitality to the warmth, shouldn't that be said about talking? For what reason does it move when it is warmed? The pinball machine's similitude is as yet worthy: think now about the pinball machine where the limits are, etc. Overwhelmingly vibrate on the grounds that they are thought about energetically.

The (electron) balls he eliminates will be out of the blue quickened in each crash, and these increasing speed charges impel electric waves. What's more, obviously, metal electrons have long queues that are especially long, the vibration of the alternate routes influences them the most, so they can't work in a get-together and communicate heat vitality uncertainly. It is obvious from such suspicions that enormous scope radiation shields are the most worthy makers.

Undoubtedly, we can be more exact: the body emanates beams at a given temperature and returns similarly as it produces similar beams.

Kirchhoff has exhibited this: the essential point is that on the off chance that we believe that a specific body can wind up superior to exchange, at that point in a room loaded with things with a similar temperature, it will include radiation from various bodies better than reestablishing vitality to them. This implies it will improve, and the remainder of the room will be cold, dismissing the second law of thermodynamics. (We can utilize such a body to construct a warm vehicle that isolates the occupying as the room gets colder and colder!)

Nonetheless, the metal sparkles when it is sufficiently warm: for what reason would it be so? As the temperature rises, the alternate way bit of the particles vibrates at a consistent level; this development scatters and accelerates electrons. Undoubtedly, even glass is lit up at temperatures sufficiently high, as electrons radiate and move.

8.2 The Absorption of the Radiation

Dark Body Spectrum

Anybody at any temperature over zero will send here and there; the vitality and recurrence of radiation rely upon a specific body structure. To start to break the warm beams, we need to state plainly about the body that does this: the most troublesome case, you can envision a sentimental body, which is the correct assurance, so likewise (from the above contention) the correct maker. Clearly, this is known as the "dim body."

Regardless, we ought to inspect our contemplations a tad: so how might we assemble the correct security? Alright, less, yet in 1859, Kirchhoff had a shrewd thought: a little hole on a huge box is a great assurance, on the grounds that any beams that discover a hole hop around within, are held near each weave, and have minimal possibility of getting out once more.

Along these lines, we can do this for a change: we have a grill with a little hole as an afterthought, and possibly the radiation from the space is adequate to show the correct producer as we will discover. Kirchhoff has aggravated researchers and experimentalists to bode well and measure (separately) the bowing/horrible intensity of this "radiation," as he calls it (in German, actually: hohlraumstrahlung, where hohlraum implies void chamber or gap, strahlung beams). Undoubtedly, it was Kirchhoff's examination in 1859 that formally eliminated quantum theory forty years afterward!

Observations

By the 1890s, arraignment techniques were progressed to the point that it couldn't be viewed as an exact estimation of the radiation's greatness in an opening, or as we would call dark radiation. In the last era of the 1800s, at the University of Berlin, Wien and Lummer twisted around the side of a totally shut oven and started to isolate the coming beams.

The exit from the opening went through the pounding of the street, which sent different waves/waves to various themes, all confronting the screen.

The locater was off the whole screen to discover how much style was communicated in each reiteration. (This is a specialist case model - genuine test blueprints are exceptionally refined. For instance, to raise hell free infrared estimations, repeating waves are executed by various quartz signals and various qualities.) They discovered rehashed radiation/twisting bends close to this (right):

The obvious range begins at about 4.3×10^{14} Hz, so this oven sparkles a dull red.

One little point: this structure is the force of the power inside the oven, which demonstrates ngo (f, T), which implies that at a temperature of T, the power of Joules/m3 in the repeat of straightforward f, f + δf is ρ (f, T) if.

To get the vitality out of the opening, recollect that the radiation inside the oven has similar waves that move in two unique ways - so half of them will come out through a hole. Also, if the hole has an A position, the waves come in during a period that will see the objective zone less. The aftereffect of these two impacts is as per the following.

Radiation Energy from the Gap Region

A = 14 A cp (f, T)

They were more set up to implement Stefan P = σT4 and Wien's Transfer Law by restricting the way that the dark body twists at various temperatures, for instance:

Imagine a scenario where we look again, and this twists in detail: discovering low waves, f, (f, T) found to compare to f2, the spellbinding state, yet by expanding f, it falls underneath parabola, fmax, at that point diminishes quickly towards zero as though the past fmax increment.

In those low waves where ρ (f, T) is spoken to, the augmentation of temperature is discovered to be twice as much radiation. Notwithstanding, what's more, in 2T, the bend follows the most redundant example before the perpetual fall - to be valid, twice as far, and fmax (2T) = 2fmax (T).

The curve ρ (f, 2T), then, is regularly the length of ρ (f, T). (See outline above.) It is also twice as wide as the even level, so the field beneath the shell, in correlation with the energetic surface, increments the temperature by multiple times: Stefan's law, P = σT4.

8.3 Basic Laws

The essential supposition that depends on the radiation test perspective on the hole.

Stefan's Law (1879)

Complete P power from one square meter of the dark region in temperature T goes as a fourth absolute temperature:

$$P = \sigma T4, \quad \sigma = 5.67 \times 10^{-8} \text{ watts/sq.m. /K4}$$

After five years, in 1884, Boltzmann found this T4 conduct in principle: he utilized conventional thermodynamic speculation for a situation stacked with electromagnetic radiation, utilizing Maxwell's wonders to relate the force and power of vitality. (The microscopic proportion of the vitality from the initial will clearly have a temperature subordinate like the radiation quality inside.) See what's happening in the notes on the nuance of the choice.

Wien Relocation Act (1893)

As the grill shifts' temperature, so does the reiteration where the radiation is sent all the more as often as possible. Actually, that is additionally legitimately identified with the general temperature:

$fmax \propto T$

(Wien himself found this law by theory in 1893, after Boltzmann's hypothesis about thermodynamic. It had as of late been watched, any place it is, similarly, by the American space expert Langley.)

Believe it or not, this skyscraper in fmax and T is normal for everybody - when the metal is warmed in a fire, the principle obvious beams (about 900K) are ruddy, almost no noticeable re-noticeable light. Further increment in T makes a hazier shade from orange-yellow, in the long run, blue to higher temperatures (10,000K or higher) when high radiation presentation is plainly noticeable.

This is a dreary advance where the best power is significant in keeping up sun-related vitality, for instance, in kindergarten. The glass should give the sun's beams access, in any case, not permit the warmth beams to come out.

This is justifiable because the two beams are at a totally extraordinary recurrence - 5700K and, state, 300K - and there are immediate to-light items that are wrong in infrared radiation. Kindergartens work in light of the fact that fmax changes with temperature.

8.4 Black Body Curve

These very much planned test outcomes are an approach to change. The essential trial of information speculating was Max Planck in 1900. He zeroed in on featuring the troublesome cases that must be available in oven partitions, which emerge from inner warmth and - in thermodynamic settings - themselves are driven by the radiation field.

Essentially, he found that he could speak to the watched bend on the likelihood that he needed these oscillators to show up as dynamic, as the good old view would ask. However, they could basically lose or take power with sections, called quanta, size hf, for augmentation oscillator f. The fixed h is at present called Planck's agreement, $h = 6.626 \times 10^{-34}$ joule/sec.

At that demand, Planck decided the extent compared to the greatness of the radiation inside the oven:

(f, T) df = 8nVf2dfc3hfehf/kT - 1

A superior comprehension of this formula with explicit tests and the ensuing requirement for imperativeness quantization turned into the most significant material science progression for a century.

In any case, nobody saw it for long! His dark body twist was totally acknowledged as a right: a developing number of direct tests demonstrated it commonly; however, the outrageous idea of quantum thinking didn't enter.

Planck didn't stress excessively - he didn't 'I trust it was conceivable, he accepted it as a unique amendment (he had trusted), after some time, which would have appeared to be ridiculous.

A contributor to the issue was that Planck's excursion to the condition was long, strenuous, and unthinkable - even to the point of making opposing suppositions of different classes, as Einstein later called attention to. In any case, the outcome was positive regardless, and to comprehend why we would follow another, easier, the course that was begun (yet not effectively finished) by King Rayleigh in England.

Chapter 9: Photoelectric Effect

Planck Radiation Law

Planck's radiation law is a numerical condition proposed by the German logician Max Planck to portray the conveyance of vitality that mirrors the energies of the dark body radiation (the assortment of thought arrives at a specific level of estimation and recaptures vitality as it ingests). Planck accepted that the cause of the radiation by the nuclear wavering and that each vibrating oscillator's intensity may have certain particular qualities, yet never have an association between them. Additionally, Planck believed that when the oscillator changed from E1 to E2, the measure of differential vitality E1 - E2 or the measure of radiation was equivalent to the result of radiofrequency, spoken to by the Greek letter - Δ and the steady h, presently known as Planck's coherence, controlled by the subtleties of the dark body radiation (e.g., E1 - E2 = hδ)

Planck's law of Eλ vitality is lit up by a unit of volume with a dark body in the scope of λ to λ + (Δλ which means increment long) can be composed by Planck steady (h), light speed (c), Boltzmann consistently (k) and complete temperature (T):

E λ = (8πhc/λ5) x (1/exp (hc/kTλ) - 1)

The length of the produced radiation compares to the mass or λ = c/v. Planck's fixed worth is characterized as 6.62607015 × 10-34 JS.

On account of the dark body, a large portion of the radiation in the electric range's infrared radiation is at a temperature of a few hundred degrees. Absolute vitality is delivered at higher temperatures, and the length of the radiation opens the change to a shorter frequency. A large portion of the radiation is discharged as noticeable light.

Einstein and Photoelectric Effect

In 1905, Einstein stretched out Planck's law to clarify the impact of photoelectric, that is, the ingestion of iron by at least one energies. The motor vitality of the electron transmitted is controlled by the μe recurrence of the radiation instead of now is the ideal time, to as far as possible μ0 at which particles can be discharged. Along these lines, when the light sparkles more, an exit happens; no evident deferral. Einstein has demonstrated that two hypotheses can clarify these impacts:

1. Light comprises corpuscles or photons, the intensity of which is given by Planck's relationship.

2. A metal particle can retain total photon or anything. Some photon vitality enters and radiates an electron that requires a fixed W vitality, metal capacity; the rest of changed over to the motor vitality of the expelled electron (meu2)/2 (I speak to the size of the electron and its speed). Consequently, the force relationship is approx.

hv = W + (meu2)/2 (1)

In the event that v is under v0, when hv0 = W, no electrons are delivered. Not the entirety of the above logical outcomes was known in 1905; however, all that Einstein said has been affirmed.

5.1 Atomic Model of Bohr

The Quantum Hypothesis was brought into the atomic field by Neil Bohr in 1913 and contributed significantly to this. Since the center of the nineteenth century, a straightforward range made of particles of power has been concentrated broadly. Low-pressure particles of iotas contain a lot of various frequencies.

This is a conspicuous difference with the power of the radiation, which spreads over a significant distance.

The frequency of various particles is known as the line range since the ingestion of beams (light) comprises of a progression of straight lines. The width of the lines is an element of the items and can make extremely complex examples. Nuclear hydrogen and antacid (e.g., lithium, sodium, and potassium) are the most straightforward spectra. On account of helium, the logical recipe characterizes the frequencies.

$1/\lambda = R\infty \ (1/m2 - 1/n2)$, (2)

At the point when m and n, the numbers n> m and R∞, usually known as Rydberg, have an estimation of 1,097373157 × 107 for every meter. With a given estimation of m, the differential lines n are arrangement. The m = 1 lines of the Lyman arrangement are in the bright part of the range; those of m = 2 of the Balmer arrangement is in the noticeable region; and those of m = 3 of the Paschen arrangement are infrared.

Bohr started with a model proposed by British researcher Ernest Rutherford, who was conceived in New Zealand. The thought depended on the investigations of Ernest Marsden and Hans Geiger, who exploded gold nuclear bombs in 1909 to the point that the bombings were completed similarly as the first gold iota authoritative; and a test by Hans Geiger, who exploded a bomb in 1909. Rutherford presumed that the molecule had a colossal stacked spine. In Rutherford's view, the molecule develops as a little close planetary system with a heart that demonstrations like the Sun as a pivoting planet-like electron.

Bohr has made three perspectives. To begin with, he contended that, in contrast to conventional material science, where there is a vast number of potential ways, the electron could be one of the circles of the purported vertical areas.

Second, he proposed that the main cycles permitted was those with an absolute number of times the intensity of the electron precise (most likely h/2т).

Third, Bohr accepted that Newton's law of motion, which controls the development of planets around the Sun, even applied to the electrons circling the core. Electron vitality (the gravitational power like the Sun and the planet) is an electrostatic fascination between a very much stacked electron and a severely charged electron. With these essential structures, he has demonstrated that the intensity of a circle has been made.

En = - E0/n2, (3)

Where E0 is a leftover centralization of the known components in, I, and, when in a steady express, the particle doesn't radiate vitality as light; nonetheless, when the electron shifts from the vitality territory of En to the province of Em vitality at low force, the measure of vitality is deducted from the recurrence v, by a given number.

We present the articulation En and utilize the relationship λv = c, where c is the most straightforward speed; Niels Bohr acquired a recipe with the specific estimation of Rydberg consistently the length of the hydrogen lines range.

5.2 Franck-Hertz Experiment

In material science, Franck-Hertz's investigation was the primary examination affirmed by James Franck and Gustav Hertz in 1914 for the presence of particular types of vitality iotas.

Franck and Hertz move low-vitality electrons to an electron tube through the gas. With the expanding electron power, some basic electron powers were found.

The electron stream has transformed from a continuous progression of gas to a full stop.

It is simply in the wake of increasing some basic points that electric iotas can utilize the vitality from electrons, demonstrating that electric particles themselves communicate out of the blue to higher obscure powers in electric molecules.

However long there is not as much as that measure of vitality in the electron assault, no change is conceivable, and the electron transition has no lit up vitality. At the point when they have a specific measure of vitality, they lose everything when they crash into electric iotas, which store vitality at a higher vitality level.

5.3 Compton Effect

To comprehend Compton's impact, Compton's appropriation ought to be followed when the two occasions are interwoven.

Compton dissemination is a steady dispersing as the frequency of diffused light changes with the developing beams, for example, gamma and x radiation photon, just as other high-power radiation. Sometimes, the radioactive aftermath of Compton happens in light of the fact that the radiation responds with the core of a particle. Be that as it may, Compton's separation regularly speaks to the cooperation between both the nuclear electron and the occasion radiation, frequently observed by the valence electron iotas from the core and less alluring, encouraging the dispersal of these electrons.

Radiation is described by an alternate frequency, called Compton move, after electron contact with radiation. A large portion of the photon vitality is moved to the scattered electrons, and the photon vitality is decreased.

At the point when the power is equivalent to the mass and the long frequency, this vitality circulation delivers more waves. This strategy is known as the Compton Effect. Compton's contrary conveyance is likewise conceivable when an electron moves a portion of its vitality to a photon.

In the twentieth century, it was imagined that when X-beams of realized radiation crashed into molecules, X-beams dispersed across edges and showed up at various statures.

In addition, as per traditional electromagnetism, which accepts that radiation demonstrations like a wave, this isn't the situation when, for instance, the θ wave signal showed after an electron contact may have a θ length. However, it doesn't. Dispersed X-beam photon is less amazing, longer, and fits underneath the episode photon.

Compton's impact is noteworthy in light of the fact that it shows that light can't be effectively deciphered as a wave. The regular perspective on electric waves can't demonstrate that there is a distinction in the width of the low-recurrence. Radiation should go about as particles to characterize the conveyance of Compton's extraordinary potential, as appeared by understanding.

Chapter 10: The Bohr-Einstein Debate

This was one of the most famous scientific conferences in history. Of the 29 candidates, 17 received or received Nobel prizes. The meeting is very important for two titans of physics: Niels Bohr and Albert Einstein.

1927 was a year, and scientists were shocked. The very existence of such an amazing thing is in jeopardy. Are electrons, lights, and similar objects, waves, or particles? In some experiments, the tiny bodies act like waves, and in others, they act like particles. This is not happening in our big world. The sound waves don't act like rocks - and luckily, your ears would bite right now.

The 1927 Quantum Mechanics conference discussed a combination of terms that seemed to be inconsistent. Schrödinger and de Broglie presented their views. But eight hundred gorillas were Bohr. It was later called the Copenhagen translation. Bohr suggested that wave measurements were defined as materials such as electrons, but as particles, organizations did not exist until someone wanted them. The act of adoption became the origin of life. Using Bohr's own words, the people involved had no "visible life in the normal context." None of that would have been Einstein.

Einstein would not have had that. The electron was an electron, and because someone was not looking at it, it was still there - wherever it "was." Towards the end of the conference, Einstein challenged Bohr's view. But that was only the beginning. By the time thirty, Einstein was dead, Bohr and Einstein were embroiled in heated negotiations - face-to-face and printing.

The conversations were of a gentleman. Bohr and Einstein were good friends and respected each other very much. However, they persisted.

He said, "It's not fair to think that physics has to find out what nature is like," Bohr said. Einstein disagreed. "The only purpose we explain to science is to find out what it is."

Bohr's definition of Copenhagen remains one of the world's most widely accepted quantum physics concepts for all its complexity. Many common definitions seem like most strangers. But all the pieces of evidence point to one simple truth. Our universe is a mystery, as all scientists will tell you. It mocks us with unimaginable facts and gives us meaning. Maybe one day, we'll go to it. But we will have to face the good mysteries around us before that.

Conclusion

The microscopic world has its own rules, which, as David Wheeler wrote, sound impossible. Some think that there must be a more reasonable and realistic understanding of the reality behind the quantum theory. One of the viewpoints for the advancement of quantum physics itself is the understanding of many universes. Wheeler says that you never know for sure until the science confirms or refutes, as with many new ideas, the latest concept convincingly. The writer says that the universe comprises not only the everyday reality but also the rest of the world, about which we learn more as science develops.

Quantum physics is generally not the first stage and may not be the last step in the continual development of our universe knowledge. It is the most progressive view of the reality of humanity for the time being. It's not just about the micro-world; it's about our daily facts. Despite this reason, Newtonian mechanics appears to be a reliable method for other practical applications. Yet then quantum mechanics will be accompanied by modern science. "Would this process be endless? Would our knowledge ever be complete? But these are questions from a different field, the field of science philosophy."

When electrons are intertwined, it means that the measurement shows the opposite of their spin signs. This interconnection occurs when the particles form in a single process. According to the exclusion principle of Pauli, every quantum system has different components. Any electron may turn out to have a positive or negative spin, but the signs are usually the opposite. The spin of one electron can be enough to automatically determine the other's spin, as in the two-slit experiment.

According to quantum theory, the second electron's spin sign is definite and opposite to the first. When one electron is measured, both electrons' wave function collapses regardless of the distance between them. The electrons demonstrate their final interconnection as part of what is known as a single quantum system. After that, electrons are no longer connected, and in the future, they will be able to acquire all properties independently. They can get entangled with new particles with which they later interact, including photons, they claim. It can continue with some delay, depending on the distance, as time and speed can never be measured with absolute accuracy.

Quantum theory proposed simultaneous interconnection would take place at any distance. Einstein denied the notion of jamming, but modern experiments showed otherwise. There are no methods for measuring time and speed with absolute accuracy, but the instruments' efficiency is improved. Scientists believe that the rate of contact approaches the speed of interconnections. Then this interaction seems to be infinite velocity, i.e., both particles at the same time acquire exact features regardless of distance (non-locality). "It is obvious why Einstein so slowly dismissed the Entanglement theory."